PROJECTS

Falkirk Council

Wood

Jen Green

Photographs by Emma Solley

WAYLAND

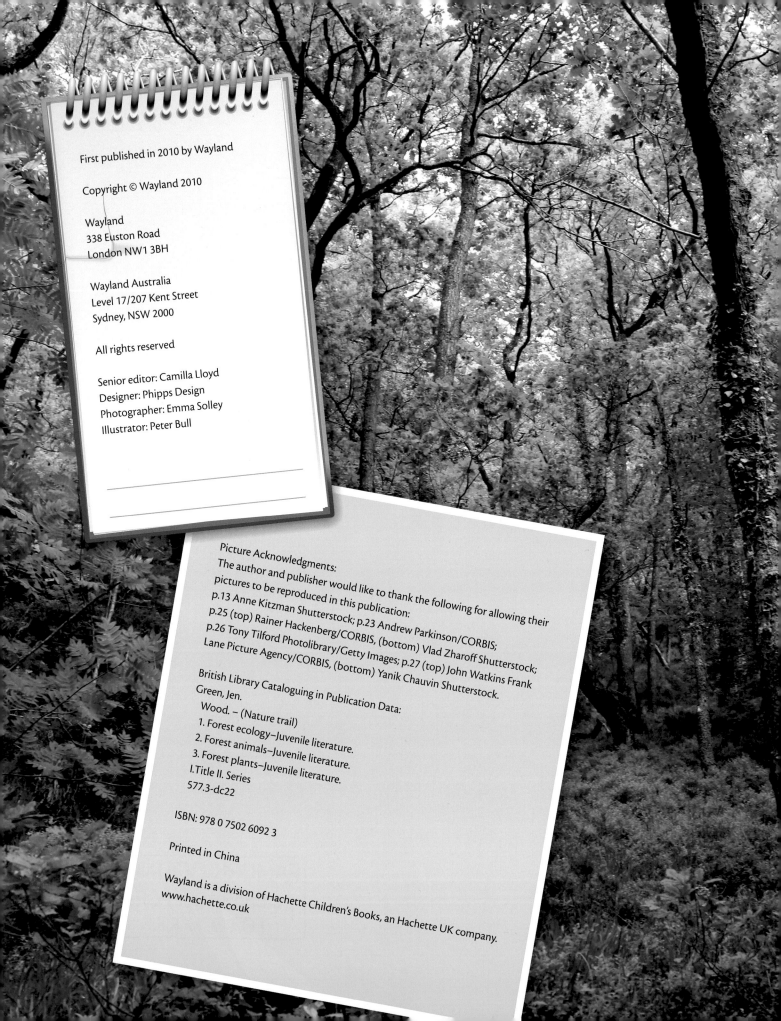

First published in 2010 by Wayland

Copyright © Wayland 2010

Wayland
338 Euston Road
London NW1 3BH

Wayland Australia
Level 17/207 Kent Street
Sydney, NSW 2000

Senior editor: Camilla Lloyd
Designer: Phipps Design
Photographer: Emma Solley
Illustrator: Peter Bull

Picture Acknowledgments:
The author and publisher would like to thank the following for allowing their
pictures to be reproduced in this publication:
p.13 Anne Kitzman Shutterstock; p.23 Andrew Parkinson/CORBIS;
p.25 (top) Rainer Hackenberg/CORBIS, (bottom) Vlad Zharoff Shutterstock;
p.26 Tony Tilford Photolibrary/Getty Images; p.27 (top) John Watkins Frank
Lane Picture Agency/CORBIS, (bottom) Yanik Chauvin Shutterstock.

British Library Cataloguing in Publication Data:
Green, Jen.
 Wood. – (Nature trail)
 1. Forest ecology–Juvenile literature.
 2. Forest animals–Juvenile literature.
 3. Forest plants–Juvenile literature.
 I.Title II. Series
 577.3-dc22

ISBN: 978 0 7502 6092 3

Printed in China

Wayland is a division of Hachette Children's Books, an Hachette UK company.
www.hachette.co.uk

Contents

In the wood

A wood is a place with many trees growing close together. Woods contain more living things than any other part of the countryside. So they are an ideal place to explore nature. There are two main types of woodlands – woods with mainly **broadleaved trees** and woods with mainly **conifer trees**.

Conifer trees such as pine, fir and spruce have narrow, waxy leaves. These trees keep their leaves all year round.

Broadleaved trees such as oak, beech and chestnut have wide, flat leaves. These trees shed their leaves for winter.

Does the wood nearest you contain mainly broadleaved trees or conifers?

Foxes, deer and birds such as this robin make their home in the wood.

This symbol warns when extra care is needed. Always take care of nature. Never pick plants. If you handle small creatures, treat them gently.

On the trail

On the nature trail you will need warm, waterproof clothing, strong shoes, a hat and sun cream.

hat

waterproof clothing

strong shoes or boots

sun cream

Make notes and drawings using a notebook, pen and coloured pencils. A magnifying glass, binoculars and a collecting jar will be useful.

collecting jar

coloured pencils

notebook and pen

binoculars

magnifying glass

Mighty trees

Trees are the most important plants in the wood. There are many types of trees, but all have the same parts: leaves, twigs, branches, a thick trunk and roots.

Each part has a vital job to do. For example, the trunk supports the tree. Roots gather moisture. Leaves use sunlight to make the tree's food.

The trunk and branches hold the leaves high in the air where they can trap sunlight.

Different types of tree have leaves of different shapes. This can help you identify the tree.

oak

beech

ash

fir

6

Leaves are like miniature factories. They use water and carbon dioxide gas from the air to make sugary food with the help of sunlight energy. These are oak leaves.

What do you see?
Study a leaf. You can see lines called **veins**, which carry food and water. Make leaf rubbings by putting paper over different leaves and rubbing the paper with a wax crayon. Leaf prints show up the veins and general shape.

stalk

vein

These are beech leaves.

Roots anchor the tree. They spread through the soil sucking up water and **minerals**.

Places to live

Trees are like multi-storey buildings. Animals and plants live at different levels in the tree: high in the leafy layer, on the trunk, on the ground or among the roots. The different levels in the tree are called storeys, just like the floors in a tall building.

Squirrels and many birds live high in trees in broadleaved forests. Conifer forests contain less wildlife.

Trees spread their leaves to make a green layer called the canopy. Birds and squirrels live here.

What do you see?

To explore the wildlife in a wood you will need to keep quiet and be patient. Take time to look at things closely. Listen for sounds and notice smells. Quick, sudden movements will scare animals away.

Insects such as this painted lady butterfly feed on flowers below the tree.

Trees are the largest living things on Earth! A big beech or fir tree can grow to over 30 metres tall.

Changing seasons

A wood changes constantly as trees and other plants adjust to the seasons. Broadleaved woodlands change more than conifer forests because these trees drop their leaves in autumn. They do this to save energy and keep in moisture.

In spring, days get longer and warmer. Buds open as trees and plants grow new leaves and blossoms. Birds build their nests and lay eggs. Animals have their young.

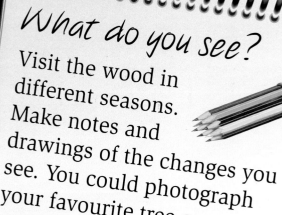

Summer brings long, warm, sunny days. Trees are leafy, and there is plenty of plant food. Animals raise their young.

What do you see?

Visit the wood in different seasons. Make notes and drawings of the changes you see. You could photograph your favourite tree or woodland scene in different seasons.

In autumn, the days get shorter and colder. Some birds fly away to warmer places. Trees produce fruits and nuts, then broadleaved trees drop their leaves.

Visit the wood in autumn. How many different colour leaves can you see?

Winter brings cold, frosty days. The wood provides shelter for animals. Broadleaved trees are bare. Animals such as dormice are asleep.

High in a tree

Birds, squirrels and hundreds of minibeasts live high in the leafy canopy. In summer, animals are hard to see among the **foliage**, but birds such as cuckoos can be identified by their songs.

Birds such as this blackbird nest in trees where their eggs and young are safe from most **predators**.

Woodland birds can be identified by their colours, actions or songs.

A jay has colourful feathers and a harsh call.

A woodpecker drums on the trunk with its beak.

A nuthatch climbs the trunk searching for insects.

A thrush has a spotted breast and a beautiful song.

Baby squirrels grow up in a nest called a **drey**.

What do you see?

Look for birds high in trees using binoculars. The best time to look is in spring and summer, when birds are busy building nests, laying eggs and rearing young.

What birds can you see among the leaves?

In the undergrowth

The ground below the tree is shaded by a leafy cover above. Plants grow in sunny patches, forming a layer called the undergrowth. Young trees sprout from seeds. These **seedlings** grow quickly towards the light.

Ferns and other plants grow in patches of sunlight.

These plants are common in woodlands.

Snowdrops bloom in late winter.

Wood anemones and primroses (below) flower in spring.

A foxglove flowers in summer – it doesn't mind the shade.

What colour flowers do you see on your nature trail?

In spring, wildflowers grow. These bluebell flowers **bloom** in April and May.

What do you see?
Make a 1-metre square on the ground using four pegs and a ball of string. Draw the plants you find in the square and try to identify them. You might need a plant book to help you.

This leaf has been nibbled by insects.

Caterpillars feed on leaves in the undergrowth.

Among the leaves

A carpet of dead leaves covers the woodland floor. Beetles, woodlice and other minibeasts feed on dead leaves and rotting animals. This helps to break down the remains and return minerals to the soil, to nourish plants. These small creatures are nature's recyclers.

The snail's coiling shell protects it from enemies and keeps it moist.

What do you see?

Study the minibeasts you find among the leaves using a magnifying glass. Counting the legs can help to identify these creatures. Insects such as beetles have six legs. Spiders have eight legs. Millipedes and woodlice have many legs. Slugs and worms have none.

Snail's eyes are on the long tentacles.

 Always wash your hands after touching soil.

Beetles are insects.
All insects have a three-part body, with a head, middle part and rear.

Antennae on the head are used for smelling and feeling.

middle section

The hard wing cases protect the delicate wings below.

rear section

Spiders trap their **prey** in their sticky webs.

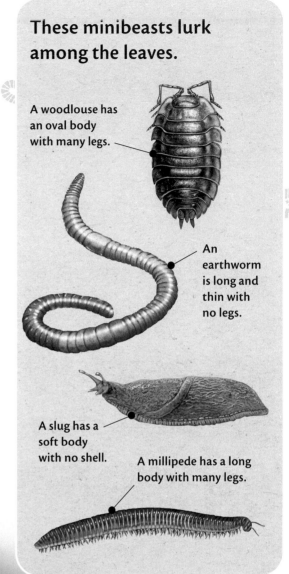

These minibeasts lurk among the leaves.

A woodlouse has an oval body with many legs.

An earthworm is long and thin with no legs.

A slug has a soft body with no shell.

A millipede has a long body with many legs.

How many different minibeasts can you find?

On a tree stump

Tree stumps and fallen trees are home to all kinds of living things that like moist, dark places. As well as plants such as moss and ferns, there are worms, slugs and other animals. Mushrooms and toadstools sprout from the stump or roots. These are not plants or animals, but a separate group of living things called **fungi**.

A tree stump is a mini-**habitat** for plants such as moss and ivy. Moss soaks up water like a sponge.

18

Ferns die back in winter and uncoil their leaves in spring.

What do you see?

Study the stump and the ground around it. Make a list of all the living things you find, grouped under these headings:

- Plants
- Animals
- Fungi

Fungi are mostly made up of fine threads that spread underground. They grow toadstools to spread seeds called **spores**.

⚠ Never touch mushrooms or toadstools, as they may be poisonous.

In a clearing

A clearing in the wood is sunny and warm in summer. The ground is grassy. Wild flowers attract bees and butterflies. Insects and reptiles such as snakes sunbathe on logs and stones.

Bees visit flowers to sip a sweet liquid called **nectar**. They also carry dusty pollen from one flower to another, which helps plants to make seeds.

These insects are found in sunny clearings.

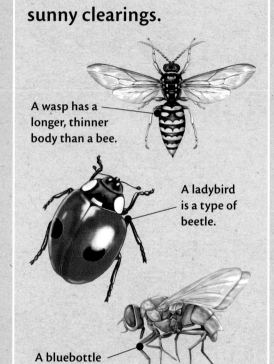

A wasp has a longer, thinner body than a bee.

A ladybird is a type of beetle.

A bluebottle is a large fly.

Snakes need to warm up in the sunshine before they go hunting. Insects sunbathe in the early morning too.

 Don't go near wasps, bees or snakes, as they could sting or bite you.

What do you see?

Butterflies suck up flower nectar using their long, straw-like mouthparts. The long tube is called a **proboscis**. If you sit quietly near flowers, you may see butterflies sipping nectar.

This butterfly, a red admiral, is sipping nectar from a flower.

How can you tell a bee from a wasp?

Fruits and seeds

Plants **reproduce** by making seeds in autumn. Many plants make their seeds inside fruits, which can be hard like an acorn, or soft like a cherry. Animals and the wind help to carry seeds far away from the parent plant, where they can sprout and grow.

Pine trees make their seeds inside cones like these.

What do you see?

Autumn is the time to look for fruits and seeds. Soft fruits include apples and blackberries. Chestnut fruits have a spiky case. Fruits with hard shells are called nuts. What fruits and seeds can you find?

Berries and seeds are eaten by birds and squirrels. The seeds pass through the bird and are dropped in another part of the wood.

These fruits and seeds are common in woods.

An acorn is the hard fruit of the oak tree.

A rose hip is a soft fruit containing seeds.

A sycamore seed floats on the wind.

Can you name the fruit of a horse chestnut which has a spiky case?

Hidden animals

Mammals that live in woods include deer, badgers, foxes, hedgehogs, stoats and squirrels. These animals are very shy and good at hiding. It can be hard to spot them, but signs such as nibbled food and footprints show that they have passed.

Grey squirrels dart among the tree tops.

What do you see?

Look for signs of mammals in your local wood, such as fur or droppings. You may find animal footprints in muddy areas.
A wildlife book will help you identify the prints. Half-eaten nuts or pine cones have been nibbled by mice, birds or squirrels.

Fallow deer have a speckled coat which blends in with the shadows. Young deer are called fawns.

Badgers leave piles of earth, stones and grass outside their burrow, called a **sett**.

What do the clues you have found tell you about the animal's way of life?

Don't touch barbed wire or animal droppings.

The wood at night

Many woodland animals sleep by day and come out to feed at dusk. Animals that are active after dark are called nocturnal. Bats, foxes, badgers, hedgehogs, stoats and owls are all nocturnal. So are moths, worms and many other minibeasts.

Night-active animals, like this bat, have senses suited to hunting in darkness.

As it gets dark in the wood, lots of nocturnal creatures come out amongst the trees and ferns.

Owls have excellent sight and hearing. They swoop low and catch prey in their sharp claws. This tawny owl is eating a worm.

What do you see?

If you go wildlife spotting at dusk, take an adult with you. Tape red tissue paper over your torch, so it won't disturb night animals. Sit quietly at a place where you have seen signs of foxes, deer or badgers. Listen out for sounds such as owls hooting or foxes barking.

The fox's pricked ears and good eyesight help it hunt at night.

Nature diary

Build up a detailed picture of life in the woods by keeping a nature diary. You could collect finds such as leaves and feathers, do a tree survey or make a map.

KEEP NOTES

Always take your notebook with you. Record the date, time, weather and place. Describe what you see. Use drawings, photos or leaves to illustrate your book.

Can you find out the name of this flower?

Date: 5 June

Time: 3pm

Weather: Sunny

Location: Undergrowth by a tree stump

Observations: Saw ants on the ground and a caterpillar on a leaf.

Make a collection of finds such as leaves, feathers, seeds, nuts and bark. Leaves and feathers can be stuck in your diary.

What insect is this? If you see a plant or animal you don't recognise, write notes, take a photo or make a sketch. Look it up later in a plant or animal book.

TOP TIPS

● Wear green or brown clothes so you blend in with the woods.

● Approach animals with the wind blowing towards you so they don't catch your scent.

● Take a camping mat so you can sit quietly on the ground.

● Don't forget to take care of nature. Take your litter home with you.

MAKE A MAP

Record all the different trees you find in a small patch of woodland. A tree book can help you identify them. Or make a map of the wood showing habitats such as clearings, streams and fallen trees.

Oak trees

Beech tree

tree stump

stream

found mushrooms here

clearing

Pine trees

Glossary

antennae The 'feelers' on an insect's head, which are used for sensing.

bloom When flowering plants come into flower. This usually happens in the spring or summer.

broadleaved tree A tree with wide, flat leaves that sheds its leaves in autumn.

conifer tree A tree that keeps its narrow leaves all year round, and produces seeds in cones.

drey A squirrel's nest.

foliage Leaves.

fungi The group of living things that includes mushrooms, toadstools and mould.

habitat The natural home of plants or animals, such as a wood or pond.

mammal An animal with hair on its body, that feeds its young on milk.

minerals The non-living materials of which rocks are made.

nectar A sugary liquid produced by flowers to attract insects.

predator An animal that hunts other animals for food.

prey An animal that is hunted by another.

proboscis The feeding tube of an insect such as a butterfly.

reproduce When plants or animals produce young.

seedling A young tree which has sprouted from a seed.

sett A badger burrow.

spore A cell that can develop into a new plant or fungus.

veins Tiny tubes in a leaf that carry water and food.

Further information

BOOKS

Look around you: Countryside
by Ruth Thomson, Wayland, 2007

**Usborne Farmyard Tales:
First Nature Book** by Minna Lacey,
Usborne, 2007

Animal Neighbours: Fox
by Michael Leach, Wayland, 2007

WEBSITES

www.bbc.co.uk/nature/animals/
The BBC's science and nature website is
packed with facts about birds,
mammals and other wildlife.

**http://feeds.bbc.co.uk/wales/wildab
outnature/explorer.shtml?Woodland**
BBC Wales's nature site has clips and
information about woodland life.

www.bbc.co.uk/springwatch/
www.bbc.co.uk/autumnwatch/
The BBC's Springwatch and
Autumnwatch sites have information
about exploring the natural world.

**www.animal.discovery.com/
animals/**
The Animal Planet website has
information, pictures and videos about
all kinds of wild animals.

Index

Nature Trail

Contents of titles in the series:

Park

978 0 7502 6094 7

In the park
Places to live
Changing seasons
In the grass
In the trees
In a bush
By the water
Under a stone
Flowers
Life cycles
Food chains
Through the day
Nature diary

Pond

978 0 7502 6091 6

At the pond
On the bank
Floating at the surface
Below the surface
On stones and plants
In deep water
In and out of water
Dabbling at the surface
Among the reeds
At the lake
Life cycles
Food chains
Nature diary

Seaside

978 0 7502 6093 0

At the seaside
Between the tides
On the cliffs
Pebble beach
Birds on the shore
Plants of the sea
In the sand
Among the dunes
In the shallows
In a shell
In a rockpool
Seaside food chains
Nature diary

Wood

978 0 7502 6092 3

In the wood
Mighty trees
Places to live
Changing seasons
High in a tree
In the undergrowth
Among the leaves
On a tree stump
In a clearing
Fruits and seeds
Hidden animals
The wood at night
Nature diary

WAYLAND